DISCARD

BRINGING BACK THE
Southern White Rhino

Ruth Daly

Crabtree Publishing Company
www.crabtreebooks.com

CRABTREE
PUBLISHING COMPANY
WWW.CRABTREEBOOKS.COM

Author: Ruth Daly

Series Research and Development: Reagan Miller

Picture Manager: Sophie Mortimer

Design Manager: Keith Davis

Editorial Director: Lindsey Lowe

Children's Publisher: Anne O'Daly

Editor: Ellen Rodger

Proofreader: Crystal Sikkens

Cover design: Margaret Amy Salter

Production coordinator and Prepress technician: Margaret Amy Salter

Print coordinator: Katherine Berti

Produced for Crabtree Publishing Company by Brown Bear Books

Photographs
(t=top, b= bottom, l=left, r=right, c=center)

Front Cover: All images from Shutterstock

Interior: Alamy: Mark Boulton 26t, Nature Picture Library 26b; Bridgeman Images: Peter Newark Americana 8; Cindy L. Croissant: 15; GaryPlayer.com: 14t; Getty Images: AFP/Carl de Souza 18, China News Service/Ji Dong 16; iStock: Blair Costelloe 23, Preez du Grobler 11, ManoAfrica 29, Catherine Withers-Clarke 27t; Nature Picture Library: Brian Stirton 22, Neil Aldridge 14-15, Anne & Steve Toon 17; New York Times: Julia Gunther 19; Public Domain: Patrick J. Fischer 6, Derek Keats 24; Seolo Africa: Guy Upfold 12; Shutterstock: Artush 10, Harry Beugelink 9, EcoPrint 7, JaySi 5b, Vaclav Sebsk 1, Thorsten Spoertein 27b, Jen Watson 5t, Catherine Withers-Clarke 4; U.S. Fish & Wildlife Service: 20; WILD Foundation: 12cr; WildAid: 21; World Rhino Day: Press & Media 28; World Wildlife Fund: 13.

Brown Bear Books has made every attempt to contact the copyright holder. If you have any information please contact licensing@brownbearbooks.co.uk

---

Library and Archives Canada Cataloguing in Publication

Title: Bringing back the southern white rhino / Ruth Daly.
Names: Daly, Ruth, 1962- author.
Series: Animals back from the brink.
Description: Series statement: Animals back from the brink | Includes index.
Identifiers: Canadiana (print) 20190233419 |
  Canadiana (ebook) 20190233427 |
  ISBN 9780778768234 (hardcover) |
  ISBN 9780778768425 (softcover) |
  ISBN 9781427124289 (HTML)
Subjects: LCSH: White rhinoceros—Juvenile literature. |
  LCSH: White rhinoceros—Conservation—Juvenile literature. |
  LCSH: Endangered species—Juvenile literature. |
  LCSH: Wildlife recovery—Juvenile literature.
Classification: LCC QL737.U63 D35 2020 | DDC j599.66/8—dc23

Library of Congress Cataloging-in-Publication Data

Names: Daly, Ruth, 1962- author.
Title: Bringing back the southern white rhino / Ruth Daly.
Description: New York : Crabtree Publishing Company, [2020] |
  Series: Animals back from the brink | Includes index.
Identifiers: LCCN 2019053203 (print) | LCCN 2019053204 (ebook) |
  ISBN 9780778768234 (hardcover) |
  ISBN 9780778768425 (paperback) |
  ISBN 9781427124289 (ebook)
Subjects: LCSH: Rhinoceroses--Conservation--Juvenile literature.
Classification: LCC QL737.U63 D35 2020 (print) |
  LCC QL737.U63 (ebook) | DDC 599.66/8--dc23
LC record available at https://lccn.loc.gov/2019053203
LC ebook record available at https://lccn.loc.gov/2019053204

Crabtree Publishing Company
www.crabtreebooks.com   1-800-387-7650

Printed in the U.S.A./022020/CG20200102

Copyright © **2020 CRABTREE PUBLISHING COMPANY**. All rights reserved. No part of this publication may be reproduced, stored in a retrieval system or be transmitted in any form or by any means, electronic, mechanical, photocopying, recording, or otherwise, without the prior written permission of Crabtree Publishing Company. In Canada: We acknowledge the financial support of the Government of Canada through the Canada Book Fund for our publishing activities.

Published in Canada
Crabtree Publishing
616 Welland Ave.
St. Catharines, Ontario
L2M 5V6

Published in the United States
Crabtree Publishing
PMB 59051
350 Fifth Avenue, 59th Floor
New York, New York 10118

Published in the United Kingdom
Crabtree Publishing
Maritime House
Basin Road North, Hove
BN41 1WR

Published in Australia
Crabtree Publishing
Unit 3–5 Currumbin Court
Capalaba
QLD 4157

Contents

Southern White Rhino Survival 4
Species at Risk 6
The Human Threat 8
Balancing the Ecosystem 10
Saving the Rhino 12
Rhinos on the Move 14
Illegal Trade and Hunting 16
Anti-Poaching Action 18
Back from the Brink! 20
Harvesting Rhino Horns 22
What Does the Future Hold? 24
Saving Other Species 26
What Can You Do to Help? 28
Learning More 30
Glossary 31
Index and About the Author 32

Southern White Rhino Survival

There are two **subspecies** of white rhinoceros in southern Africa. They are the southern white rhinoceros, also known as the southern square-lipped rhinoceros, and the much rarer northern white rhinoceros. The main threat to rhinos is from poachers. The rhinos are killed for their horns. By the 1800s, southern white rhinos were thought to be **extinct**. Then, in 1895, a small group was found in the Kwazulu-Natal region of South Africa. **Conservation** efforts have since brought the southern white rhino back from the brink of extinction. In 2011, it was described as Near Threatened (NT). Today, they number about 20,000.

There is no difference in skin color between the white rhino and other rhinos. It is thought that their name, white, is from the **Afrikaans** word "weit," which means wide and refers to their square mouths and broad upper lips, not to their color.

SIZE AND SOCIAL STRUCTURE

White rhinos are the second largest land mammals in the world, after the elephant. They have a large head supported by a muscular hump, and two horns on the end of the nose. The front horn can reach 59 inches (150 cm) in length. Adult males stand between 5 and 6 feet (1.5 and 1.8 m) tall at the shoulder, and can weigh up to 5,070 pounds (2,300 kg). Females are smaller, but can weigh around 3,750 pounds (1,700 kg). Males tend to be solitary and live on their own in a territory they mark out with piles of dung. White rhino females, or **cows**, live in groups called crashes. There may be up to 14 females and calves in a crash. Calves stay with their mothers for three years, then leave to form their own groups. White rhinos spend their days feeding and resting. In hot weather, they **wallow** in mud to keep themselves cool.

Southern white rhinos live on the savannas and grasslands of southern Africa. About 99 percent of them live across four countries. These countries are South Africa, Zimbabwe, Kenya, and Namibia. White rhinos are grazers and feed on grasses such as these pictured left in Namibia.

Species at Risk

Created in 1984, the International Union for the Conservation of Nature (IUCN) protects wildlife, plants, and natural resources around the world. Its members include about 1,400 government and nongovernmental organizations. The IUCN publishes the Red List of Threatened **Species** each year, which tells people how likely a plant or animal species is to become extinct. It began publishing the list in 1964.

SCIENTIFIC CRITERIA

The Red List, created by scientists, divides nearly 80,000 species of plants and animals into nine categories. Criteria for each category include the growth and decline of the population size of a species. They also include how many individuals within a species can breed, or have babies. In addition, scientists include information about the **habitat** of the species, such as its size and quality. These criteria allow scientists to figure out the probability of extinction facing the species.

The South China Tiger was last recorded by the IUCN in 2008. It is classed as Critically Endangered (CR) and there were thought to be only 20 individuals surviving in the wild in 2019.

IUCN LEVELS OF THREAT

The Red List uses nine categories to define the threat to a species.

Extinct (EX)	No living individuals survive.
Extinct in the Wild (EW)	Species cannot be found in its natural habitat. Exists only in captivity, in **cultivation**, or in an area that is not its natural habitat.
Critically Endangered (CR)	At extremely high risk of becoming extinct in the wild.
Endangered (EN)	At very high risk of extinction in the wild.
Vulnerable (VU)	At high risk of extinction in the wild.
Near Threatened (NT)	Likely to become threatened in the near future.
Least Concern (LC)	Widespread, abundant, or at low risk.
Data Deficient (DD)	Not enough data to make a judgment about the species.
Not Evaluated (NE)	Not yet evaluated against the criteria.

In the United States, the Endangered Species Act of 1973 was passed to protect species from possible extinction. It has its own criteria for classifying species, but they are similar to those of the IUCN. Canada introduced the Species at Risk Act in 2002. More than 530 species are protected under the act. The list of species is compiled by the Committee on the Status of Endangered Wildlife in Canada (COSEWIC).

RHINOS AT RISK

Most rhino species are classified by the IUCN as Critically Endangered (CR). In the early 1900s, the total rhino population numbered around 500,000. By 1970, the number was 70,000. Today, it is around 29,000, with the southern white rhino accounting for 20,000 of that number. One species, the western black rhino, was listed as Extinct (EX) by the IUCN in 2011. The northern white rhino is thought to be Extinct in the Wild (EW), with only two females surviving.

The Human Threat

European settlers began to arrive in South Africa in the 1600s. These **colonists** built settlements and established farms. The grasslands were plowed to grow crops. Rhinos were thought to be dangerous pests, and farmers killed them in huge numbers. Visitors to South Africa began to hunt rhinos for sport. They took the horns home as souvenirs. During the 1800s, bowls, spoons, boxes, and jewelry carved from rhino horn became big business. No one worried about the numbers of animals being killed until the 1900s, when **conservationists** took action. By then, fewer than 100 southern white rhinos remained. These were taken to live in two areas that later became the Hluhluwe-Imfolozi **Game Reserve**. There, the rhinos are protected from hunters.

> The outgoing American president Theodore Roosevelt (left) killed 11 black rhinos and 9 white rhinos on the Smithsonian-Roosevelt African expedition in March 1910. The rhinos were shipped to the National Museum of Natural History in Washington, DC.

POWER IN THE POWDER?

Poisoning was a common way to kill enemies in ancient times. In Persia (now Iran), people believed that cups carved from rhino horn could protect them from poisoning. If bubbles appeared in the cup, the liquid was said to be poisonous. The ancient Greeks believed that rhino horn had the power to purify water. It has also been used as an ingredient in traditional Chinese medicine for thousands of years, and is still used today. The horn is ground into a powder and mixed with boiling water. People believe it cures many illnesses including headaches, food poisoning, and fevers. Scientific research has proven that there is no truth in these claims, and that rhino horns do not contain any special healing qualities. The horns are actually made from a protein called keratin, which is the same substance found in human fingernails and horses' hooves. Nevertheless, in some African and Asian cultures today, rhino horns are luxury items that are thought to show how wealthy and successful a person is.

As well as hunting rhinos for their horns, humans harmed the rhino population by destroying the rhino's natural habitat to make way for towns and roads. The breakup of their habitat resulted in small rhino populations that had difficulty finding breeding mates.

Balancing the Ecosystem

Southern white rhinos help to keep the **ecosystem** healthy. When rhinos and their habitat are protected, it helps other animals and plant species to survive. A rhino can eat about 120 lbs (54 kg) of grass each day. They act like lawnmowers. This makes it easier for smaller **herbivores**, such as zebra and gazelle, to find food in the shorter grass. These grazing areas, and the pathways that rhinos make as they move through the longer grass, act as **firebreaks**. These stop fires from spreading over larger areas. When southern white rhinos were temporarily removed from Hluhluwe-Imfolozi Game Reserve, the number of fires increased.

Rhinos mark out their territory with piles of dung. Seeds are deposited in the waste, which helps to increase plant diversity in the ecosystem. The rhinos also eat any plants that may be poisonous to other animals.

THE RHINO AND THE OXPECKER

Parasites called ticks live on the skin of southern white rhinos. To get rid of them, the rhinos wallow in mud. They are also helped by tiny birds called oxpeckers, which perch on their backs. The birds peck away at the rhino's skin in search of ticks to eat. In the Swahili language, "oxpecker" means "rhino's guard." When an oxpecker spots danger, it makes a loud noise. This alerts the rhino to any nearby threats. A relationship that benefits two living things in this way is called a symbiotic relationship.

HISTORICAL RANGES OF THE WHITE RHINO

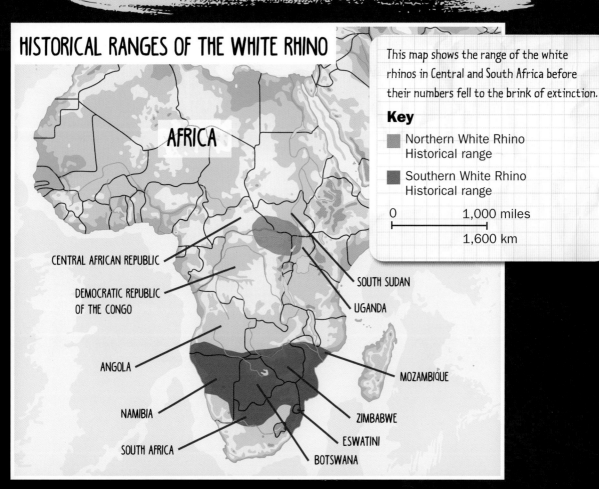

This map shows the range of the white rhinos in Central and South Africa before their numbers fell to the brink of extinction.

Key
- Northern White Rhino Historical range
- Southern White Rhino Historical range

0 — 1,000 miles
1,600 km

- CENTRAL AFRICAN REPUBLIC
- DEMOCRATIC REPUBLIC OF THE CONGO
- SOUTH SUDAN
- UGANDA
- ANGOLA
- MOZAMBIQUE
- NAMIBIA
- ZIMBABWE
- ESWATINI
- SOUTH AFRICA
- BOTSWANA

Saving the Rhino

Demand for rhino horn from Asian countries resulted in **poaching** on a large scale. By the 1950s, it was obvious that something needed to be done if the southern white rhino was to survive. Dr. Ian Player was the Senior Warden at Hluhluwe-Imfolozi Game Reserve. At the time, there was not enough space in the park for all the rhinos living there. Dr. Player came up with a plan to move some of them to Kruger National Park in South Africa. Others were sent to zoos and wildlife reserves in Africa and around the world so that populations could be established in other places. Dr. Player's conservation plan was called "Operation Rhino," and it saved the white rhino from extinction.

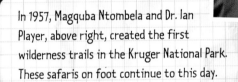

In 1957, Magquba Ntombela and Dr. Ian Player, above right, created the first wilderness trails in the Kruger National Park. These safaris on foot continue to this day.

COLLABORATING FOR A CAUSE

International trade in rhino horns has been banned since 1977. Governments, including those of South Africa, China, and Vietnam, introduced laws that made it **illegal** to buy, sell, or move rhino horns across their borders with other countries. However, it was not against the law for people to buy and sell rhino horns within South Africa until 2009. Poaching remained the biggest threat to white rhinos in South Africa. They are vulnerable because they live in herds and are not particularly aggressive. The World Wildlife Fund (WWF) is an international conservation organization. Park rangers often found themselves in danger from groups of poachers, and also wild animals, but did not always have suitable equipment. The WWF provided rangers with better equipment, and also introduced drones and other technology to protect wildlife reserves.

EDUCATING PEOPLE

In Vietnam today, many people believe that rhino horn cures cancer, even though scientists have proved this is not true. This is partly because some Asian cultures have held these beliefs for thousands of years. It is difficult for people to change their minds. This is where WildAid stepped in. WildAid is a global organization that works to save endangered species around the world. It partners with **celebrities** to help people understand the truth about the rhino horn, as well as other animal products such as elephant tusks. WildAid educates people, focusing on the simple fact that if people stop buying products made from rhino horn, rhinos will stop being killed.

Rhinos on the Move

Operation Rhino was a challenge. It was not easy to catch and move southern white rhinos because of their speed and size. Although they usually move slowly, rhinos can run as fast as 30 miles per hour (48 kph). To catch the rhinos, park rangers used dart guns containing anesthetic, which made the rhinos sleepy and wobbly on their legs. The rangers followed them on horseback until they were tired enough to capture safely. The rhinos were guided into crates, where they were tagged to help with identification. The captured rhinos were kept at a base camp for about three weeks to check on their health, before being moved to other locations in South Africa. In the 1960s, Operation Rhino **translocated** 4,000 southern white rhinos, including 350 that were sent to the Kruger National Park.

Dr. Ian Player measures a southern white rhino during Operation Rhino, in the 1960s, left. Today, rhinos continue to be moved for their protection. Below, conservationists and vets release a white rhino from South Africa into a new park in Botswana.

COLLABORATING FOR A CAUSE

When poachers or smugglers are caught carrying rhino horn, they often claim they are innocent, and that they do not know where the rhino horn has come from. In the past, lack of evidence has made it difficult to convict them of their crimes. However, in 2014, the WWF set up a rhino database to cover all of Africa. This is the Rhino DNA Index System (RhODIS). In South Africa, researchers and law enforcement officers were trained to collect DNA from the rhinos and the areas where poaching had taken place. This information was entered into the database to find matches. Other countries, including Kenya, Botswana, and Zimbabwe, began to use RhODIS, which makes it easier to identify and convict smugglers and poachers successfully. The DNA database provides vital information that links rhino horns to the rhino, the poaching site, and the poacher.

SUCCESS STORY

The San Diego Zoo Safari Park received 20 southern white rhinos from South Africa in 1971. Over the years, more than 90 calves have been born from these rhinos. Justin (right), one of the youngest calves, was born in February 2018. Many rhinos have since been moved from San Diego to other zoos, enabling the southern white rhino population to increase around the world.

Illegal Trade and Hunting

By 1977, the trade in illegal rhino horns had become such a serious problem that the Convention on International Trade in **Endangered** Species of Wild Fauna and Flora (CITES) officially banned it. This meant that it was against international law to buy or sell rhino horns or any products made from them. Vietnam introduced tough prison sentences for people found guilty of smuggling rhino horn. In Mozambique, park rangers, border guards, local police, and wildlife officers began to work together to arrest poachers. Although these laws made it more difficult for people to buy or sell products made from rhino horns, trading, smuggling, and poaching continued. Rhino horn is worth more than gold. The horn of an average **bull** rhino weighs approximately 22 lbs (10 kg). In Asia, this could be worth $15,000 to $30,000 per pound.

A customs officer in Guangzhou, China, displays products made from ivory and rhino horns. In just four months, the anti-smuggling bureau of Guangzhou Customs seized 393 lbs (178 kg) of smuggled horns and other animal products.

TROPHY HUNTING

Many southern white rhinos are kept on private farms. Trophy hunting is organized on these farms for people who hunt big game animals, such as rhinos, for sport. Trophy hunting is legal in Namibia and South Africa, but many African countries do not allow it. Trophy hunters are allowed to keep the rhino's head and horn as a souvenir. This is preserved by a **taxidermist** and shipped to the home of the trophy hunter for display.

Sometimes there are too many bull rhinos on the farm, or some bulls are more aggressive than others and fight with the other rhinos in the crash. Farmers allow these bulls to be hunted so they can protect the rest of their rhinos. Hunting permits usually cost about $100,000, but they can be much more. The money helps farmers to pay for anti-poaching security, such as armed patrols and electric fences. However, more rhinos are still killed by poachers than trophy hunters. Between 2012 and 2017, 90 rhinos were killed by hunters and 1,140 rhinos were killed by poachers. Some rhino farmers sell their rhinos because the cost of protecting them from poachers is too great.

Anti-Poaching Action

Technology is important in the fight against poachers. Thermal cameras, sensors, and global positioning systems (GPS) **monitor** and track people behaving suspiciously in or near wildlife parks. Parks have secure fences and armed patrols. Anti-poaching response teams search areas on foot, on horseback, in trucks, and by helicopter. Drones are used to send information to anti-poaching patrols, and some drones are fitted with night vision cameras. Rangers patrolling some of the most dangerous areas are equipped with binoculars, camping equipment, radios, and body armor. Rangers also use tracker dogs, which can smell a person over a distance of about 0.5 miles (0.8 km). Dogs are used not only to detect poachers, but also to uncover weapons that may have been hidden by the poachers.

> The tactical anti-poaching response team is shown here at the Ol Pejeta **Sanctuary** in Laikipia, Kenya, in 2016. Ol Pejeta is home to the world's last two northern white rhinos.

COLLABORATING FOR A CAUSE

The Black Mambas are the only all-women anti-poaching group in the world. Members of the unit come from the communities that border the Balule nature reserve in South Africa. They aim to stop the poachers before they can kill the rhinos. The Black Mambas patrol the park borders and destroy poachers' camps. They also stop traffic to search for any rhino products being **smuggled** out of the area. Since the unit was formed in 2013, approximately 75 percent fewer rhinos have been poached from the Kruger National Park. The unit also started the "Bush Babies" project to educate schoolchildren about rhino conservation.

HELPING LOCAL COMMUNITIES

Many people living near wildlife parks are poor and unemployed. They have no running water or electricity. Villagers are sometimes involved with rhino poaching because the money they earn improves the quality of their lives. Local villagers do not benefit from money paid by tourists and hunters, so they turn to poaching. Some conservation organizations are trying to change this situation. At Hluhluwe-Imfolozi Game Reserve, tourists pay a conservation fee that is used to help villages nearby. Some villagers are employed at the reserve. Others sell tourists crafts or produce from their farms. Many rhino ranches have tourism programs that employ local people as guides on treks, bird-watching trips, and photography tours. When local people understand the benefits the rhinos can bring, they are less likely to become involved in poaching activities.

Back from the Brink!

Today, there are approximately 20,000 southern white rhinos living mostly in South Africa. There are smaller populations in Kenya, Namibia, and Zimbabwe. In South Africa, rhino poaching is at its lowest for five years, but it is still a serious threat. Rhinos have to be kept in National Parks or private game reserves where they can be protected. Secure fences, effective monitoring, and armed anti-poaching patrols have all made life more difficult for poachers. An important factor in bringing the southern white rhino back from the brink of extinction has been the establishment of rhino populations at zoos around the world.

A ranger in Nakuru National Park in Kenya closely monitors the health of the rhinos under his protection.

WILDAID "NAIL BITERS" CAMPAIGN

Rhino horns and human fingernails are made from a protein called keratin, which forms strong skin, nails, and hair. But keratin cannot cure cancer. In some cases, it has been linked with causing cancers. In 2016, WildAid launched their "Nail Biters" campaign aimed at the illegal rhino horn trade in China. Rhino horn is widely used in traditional Chinese medicine. Celebrities took part in a campaign in China to inform people about rhino horn. They were photographed biting their nails, with explanations written in English and Chinese. These photographs were put on billboards in China's major cities, such as Beijing and Shanghai.

COLLABORATING FOR A CAUSE

Many countries worked together to save the southern white rhino from extinction. In the United States, the U.S. Fish and Wildlife Service (USFWS) supports conservation programs in countries with southern white rhino populations. It funds anti-poaching programs, research, and education. The USFWS also works with some African governments to bring justice to wildlife smugglers. In 2019, a rhino smuggler was arrested in Uganda and was successfully prosecuted in New York City for his crimes.

Harvesting Rhino Horns

Rhinos sometimes break or damage their horns, but they will grow back. Some conservationists think that harvesting rhino horns helps to reduce poaching, but not everyone agrees. Harvesting, or dehorning, means that rhinos have their horns removed **humanely** by a **veterinarian**. This was first used to discourage poaching in Namibia in the 1980s. Dehorning usually takes about 15 minutes. The rhino is **sedated** and its eyes and ears are covered. When the horn has been removed, a small stump remains, which is coated with tar to stop it from becoming dry or cracked. The horn eventually regrows from the stump. Many rhino farmers have large stores of harvested rhino horns. Some people say they should be sold to pay for conservation projects, but other experts think that if they were sold legally it would lead to an increase in poaching.

A veterinarian cuts the horn from a sedated white rhino at a game farm in North West Province, South Africa.

REASONS FOR DEHORNING

- **Dehorned** rhinos are of no value to poachers.
- Rhino horn grows back within 18 months to 2 years. They can be removed again without harming the animal.
- Money from the sale of rhino horns could be used to pay for the costs of protecting rhinos. Some rhino farmers argue that if they cannot sell the horns, they cannot afford to continue protecting their rhinos. Keeping areas secure from poachers can cost over $300,000 per month.
- Demand for rhino horn in Asia is still high. The use of rhino horn in traditional medicine gives false hope to sick people.
- Dehorned rhinos are kept together and protected.

REASONS AGAINST DEHORNING

- Poachers will continue to kill dehorned rhinos because even the small horn stump is valuable. In Save Valley Conservancy, Zimbabwe, some rhinos were killed for their stumps just days after being dehorned.
- Rhino horns sold in South Africa could be smuggled to other countries.
- A rhino uses its horn for many things, including defending itself, guiding calves, and digging for water.
- Storing and transporting rhino horns is difficult. They must be kept in highly secure places so that they are not stolen.
- Any kind of surgery involves some risk to the animal.

What Does the Future Hold?

More than 80 percent of white rhinos live in South Africa. Between 2013 and 2017, more than 1,000 white rhinos were killed by poachers each year. Although action against poaching has had some success, it continues to be a major problem. Criminal gangs can become wealthy very quickly. In 2018, customs officers in Malaysia discovered 50 rhino horns being shipped to Vietnam. Their value was approximately $12 million. In 2019, officials in Vietnam discovered more than 275 pounds (125 kg) of rhino horn hidden in blocks of plaster. Other horn is smuggled in the form of beads, bracelets, and bangles, which makes it difficult for border guards to detect.

Climate change may also impact white rhinos in the future. During a drought in South Africa in 2015 and 2016, the southern white rhino population suffered. There was not enough grass for them to eat. They also need water to wallow in and keep cool, and to remove parasites from their skin.

CHANGING LAWS

Some rhino farmers in South Africa were successful in changing the law so that rhino horns could be sold within the country if they had been dehorned humanely. Since 2017, the trade of rhino horns within South Africa has been legal. It is still against international law for trading between countries, although the governments of Eswatini and Namibia proposed changes to this at the CITES conference in 2019. Both countries wanted international trade of white rhino horns to be legalized, but only for horns that had been removed legally. Namibia also proposed making it legal for live rhinos to be sold to other countries. However, CITES did not accept these proposals. A plan by China in 2018 to lift the ban on the use of rhino horn in medicines was also postponed following pressure from conservation groups.

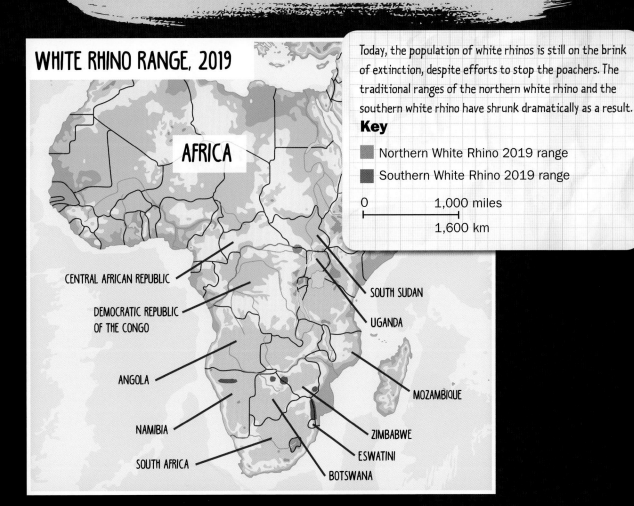

Saving Other Species

Lessons learned in the fight to save the southern white rhino are being used to protect other rhino. There are five rhino species: White, Black, Javan, Sumatran, and Greater One-Horned. The threats faced by all rhino species are similar, and it is possible that many of the methods used to save the southern white rhino will prove equally successful. However, it is too late for the western black rhino subspecies, which was declared extinct in 2011.

THE NORTHERN WHITE RHINO

In 2009, four of the world's last surviving northern white rhinos were relocated from captivity back to the wild in a last bid to save the species.

Two males, Sudan (left) and Suni, and two females, Fatu and Najin, were transferred from a zoo in the Czech Republic to the Ol Pejeta Conservancy in Laikipia, Kenya. It was hoped they would mate and start a new population. Guards protected them from poachers. Unfortunately, both males died from old age, Suni in 2014, and Sudan in 2018. Fatu and Najin are the only two northern white rhinos left in the world. They are classified as Critically Endangered (CR), Possibly Extinct in the Wild (EW), on the IUCN Red List.

THE BLACK RHINO

The black rhino lives in sub-Saharan Africa. One hundred years ago, they numbered around 100,000, but by 1970 this number had fallen to 65,000. The numbers continued to decline due to hunting and poaching. By 1995, there were only 2,000. However, conservation efforts have rebuilt populations to around 5,000 today. The black rhino is classified as being Critically Endangered on the IUCN Red List.

RHINOS IN ASIA

There are approximately 60 Javan rhinos remaining in the world. All live in the protected area of Ujung Kulon National Park in West Java, Indonesia. Their skin has thick folds and they only have one horn, with some females not having a horn at all. The Javan rhino is the rarest of all rhino species and is classified as Critically Endangered (CR) on the IUCN Red List. The Sumatran rhino (CR) is the only Asian rhino with two horns. Sumatran rhinos number less than 80, and are also classed as Critically Endangered on the IUCN Red List. The greater one-horned rhino (right) is found in India and Nepal. Thanks to conservation protection, the population is now over 3,500. It is classified as Vulnerable (VU) on the IUCN Red List.

What Can You Do to Help?

The southern white rhino is back from the brink of extinction, but it is far from safe. Saving rhinos takes time and commitment from dedicated groups of brave conservationists and park rangers. Although Africa is far away, actions by people all over the world, including in North America, can help to save these animals before it is too late.

Organizations such as WWF, Save the Rhino, Saving the Survivors, Care for Wild Rhino Sanctuary, and the International Rhino Foundation are actively involved in conservation and research. They also help to save the survivors of poaching attacks. Raising whatever money you can with events, including bake sales and car washes, will help in the fight against poachers.

World Rhino Day was first announced by WWF South Africa in 2010. Zoos, rhino sanctuaries, and people across the world raise awareness and vital funds for the five species of rhino that are all under threat from the illegal rhino horn trade.

If you are lucky enough to visit Africa on a vacation, be sure that you never buy illegal wildlife souvenirs. In particular, never buy anything made from elephant tusks or rhino horns. If you are not sure, don't buy it.

SPREAD THE WORD

You may not be able to travel to see rhinos, but you can learn more about them. Spreading information and educating people is a great way to help. Here are some ideas for things you could do:

- World Rhino Day is held on September 22 every year. Find out what your local zoo is doing and join in with some activities, or why not get your school or community to host an event on World Rhino Day?

- Write your elected representative or local newspaper and ask them to support the international groups working in the fight against wildlife crime.

- Prepare a PowerPoint presentation to show to your class explaining how the southern white rhino came back from the brink.

- Write a story or design a picture book story about the southern white rhino. Ask your teacher if you can display it in your school library.

Learning More

Books

John W. Kimball Learning Center, The Fifth Graders of P.S. 107. *One Special Rhino: The Story of Andatu*. Beast Relief, 2014.

Katirgis, Jane, and Jan M. Czech. *Endangered Rhinos*. Enslow Publishing Inc, 2016.

Markle, Sandra. *The Great Rhino Rescue: Saving the Southern White Rhinos*. Millbrook Press, Lerner Publishing Group, 2018.

Meeker, Clare Hodgson. *Rhino Rescue!* National Geographic Kids, 2017.

On the Web

www.youtube.com/watch?v=y2poHlCOOco&list=FLO6xU4MTuB7 sKtfa1vhumWw&index=1&spfreload=10
This video takes you back in time to see the early days of Operation Rhino in action.

kids.sandiegozoo.org/animals/rhinoceros
This San Diego Zoo site has some interesting information about rhinos. Plus, hear a rhino bellow!

fossilrim.org/animals/southern-white-rhinoceros/
Pictures and facts about the southern white rhino.

www.youtube.com/watch?v=ht-OthPgu78
Watch a short video featuring Danai Gurira as part of WildAid's rhino campaign.

www.savetherhino.org/rhino-info/
The Save the Rhino website gives an overview of the five rhino species, including population figures, threats, and what you can do to help.

senecaparkzoo.org/rhino-conservation-story/
The Seneca Park Zoo is a partner with the International Rhino Foundation. This page summarizes the conservation story of the southern white rhino.

Glossary

Afrikaans A South African language based on the form of Dutch introduced to the country by settlers in the 1600s

bull An adult male rhino

celebrities People who are well-known and popular locally, nationally, or around the world

colonists People from one country who live in and control another country

conservation The preserving and protecting of plants, animals, and natural resources

conservationists People who work together to protect plants, animals, and natural resources

cows Adult female rhinos

cultivation Animals or plants that are raised or grown with human help

dehorned An animal, such as a rhino, that has had its horn removed

ecosystem Everything in a particular environment, including animals, plants, and nonliving things such as soil

endangered In danger of becoming extinct

extinct Describes a situation in which all members of a species have died, so the species no longer exists

firebreaks Areas of land that are cleared to stop a fire from spreading

game reserve A protected area of land for large mammals

habitat The natural surroundings in which an animal lives

herbivores Animals that eat plants

humanely In a way that is kind and causes minimal pain

illegal Against the law

monitor To observe closely and record information regularly

parasites Organisms that live on or inside living things, usually harming their hosts

poaching The illegal hunting and killing of animals

sanctuary A safe area

sedated An animal that has been made calm and quiet with a drug

smuggled Something that has been moved from one place to another in secret, usually an illegal activity

species A group of similar animals or plants that can breed with one another

subspecies A smaller group within a species that shares similar characteristics but is its own animal

taxidermist A person who prepares dead animals so that they have a life-like appearance

translocated To move something to another location

veterinarian A person trained and qualified to treat diseased or injured animals

wallow To roll in mud

Index and About the Author

A B C
anti-poaching response teams 18
Black Mambas, The 19
black rhino 7, 8, 26, 27
Botswana 11, 14, 15, 25
China 6, 13, 16, 21, 25
Chinese medicine 9, 21

D E H
dehorning 22, 23
DNA 15
ecosystem 10, 31
habitat 6, 7, 9, 10, 31
historical range 11
Hluhluwe-Imfolozi Game Reserve 8, 10, 12, 19

I J K N
Illegal trade 16, 17
Justin the baby rhino 15
Kenya 5, 15, 18, 20, 26
Kruger National Park 12, 14, 19
Kwazulu-Natal 4
Namibia 5, 11, 17, 20, 22, 25
National Museum of Natural History 8
northern white rhino 4, 7, 11, 18, 25, 26
Ntombela, Magquba 12

O P
Ol Pejeta 18, 26
Operation Rhino 12, 14, 30
Oxpecker 11
Player, Dr. Ian 12, 14
poachers 13, 15, 16, 17, 18, 19, 20, 23, 24, 25, 26, 28

R S T
rangers 13, 14, 16, 18, 28
Rhino DNA Index System (RhODIS) 15
rhino farmers 17, 22, 23, 25
rhino horn 8, 9, 12, 13, 15, 16, 21, 22, 23, 24, 25, 28, 29
San Diego Zoo 15, 30
smuggling 16
South Africa 4, 5, 8, 11, 12, 13, 14, 15, 17, 19, 20, 22, 23, 24, 25, 28
South China Tiger 6
trophy hunting 17

W Z
western black rhino 7, 26
WildAid 13, 21, 30
World Rhino Day 28, 29
World Wildlife Fund 13
Zimbabwe 5, 11, 15, 20, 23, 25

ABOUT THE AUTHOR
Ruth Daly has over 25 years teaching experience, mainly in elementary schools, and she currently teaches Grade 3. She has written more than 45 non-fiction books for the education market on a wide range of subjects and for a variety of age groups. These include books on animals, life cycles, and the natural environment. Her fiction and poetry have been published in magazines and literary journals. She enjoys travel, reading, and photography, particularly nature and wildlife.